TERRI WOLFE FENNER

MATH WITHOUT RULES

Crossing the Bridge to Understanding
Volume I

This book is the first in a series to be written by
Terri Wolfe Fenner. The main purpose of this series
is to provide guidance to students, parents and
educators in the area of mathematics.

Math Without Rules

Terri Wolfe Fenner

This book is dedicated to:

My Parents, Bill and Jeanne Wolfe, whose example as educators lead me down the right career path and whose love and guidance provided me with everything I ever needed.

My husband, Gregg and two sons, Noah and Jonah, who believe in me and always provide me with love and support.

My High School Mathematics teacher, John Vergne, who instilled in me the love of math and inspired me to become a high school math teacher.

Introduction

Mathematics? This word has several meanings depending on who is answering the question. To some it is a means of balancing a checkbook, to another it is a way of making a living such as an accountant, some see it as a means of building the largest bridge needed to cross a body of water but to most it is a dreaded course that one is required to take all of the way throughout their time in school. One of the main goals of this book is to present many of the essential concepts of mathematics in a way different than most books would do so and to demonstrate various ways of comprehending the concepts.

Throughout my numerous years of teaching mathematics, I have taught students of various ages and levels of abilities. During that time, I have realized that not all students can learn the same material in the same way. Experimenting with various techniques and strategies of my own, I now can present mathematics to others in more than one way. That means I definitely vary from the traditional math book!

Even though many ideas and concepts will be presented in this book for particular grade levels, it is important to remember that mathematics is a subject that builds on previous knowledge. Thus the information can be utilized at any level based on the ability of the child. Visualizing mathematics as an interstate with many ramps entering the main road at different places along the route will enable individuals to understand that all sections of the subject are leading to the same point at the end of the journey.

As the reader utilizes this book, I will stress many important concepts such as the use of a dictionary, the importance of alphabetical order, Mrs. Fenner's Golden Rule of Math and most of all working on learning the subject as one big area instead of separate individual sections.

Individual Notes

<u>Individual Notes</u>

Chapter 1

This chapter will be the beginning of the journey of learning mathematics through new and creative ways. The information shared in this chapter is not just for a particular level or a particular class, but pertains to ideas that can be utilized throughout one's mathematics career.

The first idea is that of using a traditional dictionary when learning mathematics. Today it is quite evident that the use of the computer in most cases has replaced this important tool. Whether working with one or several individuals, using a dictionary brings a hands on experience to the lesson. Many of the terms used in mathematics are quite similar to common words daily used by most people. For example if teaching the "Identity Property of Addition" have the individual look up the word "identical". Of course the word means to be exactly the same. When one looks at an example of this property, it is easy to see that the equation starts and ends with the same "identical" value. Ex. $8 + 0 = 8$. Building connections between vocabulary that one already knows to that which is being used in mathematics is not exactly teaching a new concept but just transferring previous knowledge to a new area.

Another important use of the actual dictionary is to reinforce the concept of reading across the curriculum. When students utilize the dictionary, many earlier learned ideas are reinforced such as alphabetical order, noting if the word is a noun or verb and most of all it is possible for them to read surrounding words and learn their meanings as well. It is impossible to emphasize all of the numerous uses that a dictionary can provide to students both in and out of math class.

Throughout all the years of teaching mathematics, the realization that a concept taught in Kindergarten is found quite frequently in many areas of the subject. That concept would be that of "Alphabetical Order". My personal ABC'S as a teacher consist of making the math Achievable, Believable and Conceivable. After teaching the area of shapes using this idea, an awareness evolved to look for it in other formulas and characteristics of mathematics. Some examples of using alphabetical order includes: **B**ase times **H**eight, **L**ength times **W**idth and ordered pairs **(X , Y).** Later examples throughout the book will demonstrate several more uses of this idea of teaching in alphabetical order.

Another concept that is easy for students to understand is what is called "Mrs. Fenner's Golden Rule of Mathematics". This rule is used to solve basically any mathematical problem involving finding the value of a variable. The basis for the rule is that in order to begin solving the problem one must start with the value that is located on the side of the equation by itself. Secondly, the student must use the "Opposite" function to solve the problem. This rule works no matter which number is larger or smaller, which one is listed first or last, and most importantly it works for various computational problems (addition, subtraction, multiplication, division, percents, algebraic equations, etc.). The rule can be taught at any level of learning and any grade level. Let's look at an example of this rule. The equation is $5 + X = 20$. It is simple to compute that this answer would be 15 but take a minute and think about why. Most would say that it was done by starting with 20 and subtracting the 5. Did the rule apply to this question? YES! The number 20 was on the right side of the equal sign by itself where on the left side one finds a 5, + sign and an X. The opposite function was used to solve the equation. Another example would be $9 = x/3$. The answer would be found by starting with 9 and multiplying by 3. Once again the rule applies to the problem. Further in the book, it will be obvious that this rule always works and is easy to remember. Just think about it, if someone (or in the case of a math problem, the value) is the only one there then they have to go first because there is no one (or value) to go before them.

Finally, as one reads through the book it will be quite obvious to see that these previously mentioned concepts will be found at all levels of mathematics and can be utilized for many areas as well. This is one attempt to demonstrate that it is important to bring together as many mathematical ideas as possible, into as many problems as possible, in order to continually practice many concepts over and over again.

<u>Individual Notes</u>

Individual Notes

Individual Notes

Chapter 2

As most people know, the majority of what one knows now was supposedly learned in Kindergarten. The learning years from Kindergarten through Second Grade is also crucial in the development of a student in the area of mathematics. Throughout this time in a person's life, the strong foundation must be built for the continual mathematical process to start.

During this time children develop what I call "One's Attitude Of Mathematics". It is important for both the educator and the parents to emphasize that mathematics is a concept that is essential for the rest of their lives. At this stage, the children must understand that math is and will be used on an everyday basis. Be sure to instill and demonstrate how math directly relates to various activities such as telling time, cooking, MONEY, distance (Have you ever heard "Are we there yet?") and even counting how many times one must rinse their hair when shampooing. I encourage you to incorporate math into their lives whenever possible.

At this stage, I believe there are three basic numbers to concentrate on at the beginning. Those numbers are 2,5 and 10. There are several reasons for selecting these numbers but let me just give you a few of my own personal ones. First of all I select the two because students realize that they have two hands, two feet, must wear two shoes and usually in Kindergarten one must always have a partner. There is a lot in their lives right now that can be associated with the number 2. Secondly, I focus hard upon the number five. Five represents that number of fingers on each hand, tally marks are always in groups of five, money amounts are closely associated with the counting of five and another important idea is that of telling time in which the minutes are in increments of five on the clock. Finally the number ten is also important at this age. Some examples of this number are the same as that of the five except in larger groups such as tally marks, the minutes on a clock such as 10 after, 5:30, etc and $1.50 which is made up of 15 dimes (10 cents).

After instilling the importance of mathematics and utilizing the three basic numbers of 2,5 and 10, one can introduce the concept of distance and size. This can sometimes be a difficult concept for children at this age but using visual tools and objects that they are familiar with will be a huge advantage. Let's start with the concept of distance and what visuals can be used with children at this age. The units that are going to be introduced are inch, feet, and yard. Students are not going to be responsible for converting from one unit to the other but could be asked which unit to utilize to measure a specific item.

The three items I chose to use all have the word of the unit inside the illustration. For example, for inch I would use the visual of an "inch worm". The inch worm is a very small worm in which the student can visual through the use of a picture of one or even a live one that the unit inch would be used to measure small objects. The next measurement which would be "feet" and of course one would use the concept of a foot. (Here is also an opportunity to introduce a "reading" idea that feet is the plural of the word foot.) You would want to use your foot as the example since many of these young students probably still have a rather small one. To illustrate the difference between the inch and the foot using a hands on activity, have one group of students mark with a small line the beginning and end of the inch worm on a piece of paper while the others do the same except using the teacher's foot. Then have the students complete the drawing by connecting the ends of their lines (which will make somewhat of rectangular shape) to make the shape of a ruler. In conclusion discuss the concept of "yard" and where they have heard that word before. Of course many will talk about that green space that they play in at home. Ask them if it is smaller or larger than the inch worm and do the same for the word foot. Once they have answered correctly, discuss that the inch is used to measure small objects, the foot is used for medium objects and the yard is for larger ones.

At this level, it is also important to start the concept of addition and subtraction. Even though you may choose to use simpler vocabulary such as "put with", "make bigger", "add to", "take away from", "make smaller" or others, the idea is being introduced at a younger age. This is such an easy concept to make visual for students because one can actually utilize objects for the students to count.

In the Kindergarten years the emphasis is placed upon counting and sorting but subliminally one is introducing the addition and subtraction concept. In first grade, it is important to stress the words such as plus, add, minus and subtract but to also remain using the ones that were used during the previous year. Also introducing the concept of how to write an equation using the visual items will directly lead to the math facts. During the second grade year then it should be an easy transition for students to understand why 2+3=5, because they have spent two years visualizing the equation through images. Many times students (at any age) will comprehend the concept more easily when the understanding is demonstrated of where it came from and why it works.

<u>Individual Notes</u>

Individual Notes

Individual Notes

Chapter 3

As a mathematics educator, personally I view the next set of grade levels (3-5) of utmost importance. During these years, the information that students are presented is so vast that it may seem at times that there is no possible way to cover all of it.

Throughout these grade levels, educators should develop and use what I call "Power Lessons". These are lessons that are geared to teach more than one concept per lesson. As this chapter progresses, this type of lesson will appear throughout. This concept is also another way of bringing various mathematical ideas together and demonstrating how this whole concept works together.

During this time, it is very important for students to understand the processes of multiplication and division. In many instances, students are forced to memorize the multiplication and division facts (which I believe must take place). But I also believe that students most of the time view these facts as something else they are just being forced to do. One way of introducing these facts to students is to demonstrate to them that these facts are nothing more than a different form of writing an addition or subtraction problem. Now, many people ask me the question "How can that be? They look nothing alike!" True, they may look nothing alike in equation form but they are very similar in theory form.

Throughout my teaching career, I have definitely realized that the way the material is presented directly relates to the final outcomes. So let's take a minute and think about how addition is very similar to multiplication. If a student has 2 groups with 5 apples in each group, the addition equation would be 5 apples + 5 apples = 10 apples. The student would be taught that it is possible to put the 2 groups together and get a total of 10.

Now when one switches to the concept of multiplication, it is no longer adding apples to apples but multiplying the number of GROUPS by the number of APPLES in each group. This is a great opportunity to introduce the concept that the word "of" means to multiply and "is" means equal in math. Have the students take the problem and write it as follows: 2 Groups OF 5 Apples IS ? (Now the students are also practicing how to write Extended Response Answers by converting math problems to sentences and visa versa.) My question to you is "Do you see a "Power Lesson" evolving in this lesson?". Ask yourself these questions: "How many different ideas have I covered so far in this lesson?"; "Have I reviewed something that they already know?; "Have the students learned any new mathematical terms?;

and "Have I assisted them in being able to see that mathematics is a progression of taking what you already know and putting it in a different form?". Students need to be able to see that something they learned in the PAST is assisting them with what they are doing NOW.

Lets switch gears and talk a little about division. This seems to be one of the hardest concepts for students in grades 3-5 to understand. Whether it is a basic division fact or especially "long division", students always seem to struggle with this area. Here is a great time to use the DICTIONARY!! Have the students look up the word "divide". The main word that students will come up with is "separate". So in mathematics the word divide must mean to separate a number into parts. Now it is important to talk about that math is a subject that is always "FAIR". Referring back to the previously mentioned problem with the apples, if one apple is taken from one group then one must be taken from every group. Now lets think about how one can show that dividing or separating in mathematics must be "FAIR" to these younger students. Say that you take a bag of candy into the classroom and you are going to "divide"it among your students. In order to be "FAIR", each student must get the same amount of pieces of candy. Using the apple question, if one has 10 apples and 2 groups of students, in order to be "FAIR" how many apples would go to each group?

Next lets look at the dictionary again and look up the word "remainder". Spend a little time and discuss "synonyms". One of the synonyms for remainder is "leftover". That is exactly what a remainder is in mathematics. The remainder is the number of the items leftover when there is not enough to distribute to each group again.

Long Division is another area that many students struggle with during this time in mathematics. I believe that the students understand the concept but quickly become discouraged when they have to keep guessing throughout the process. I know that when I learned this procedure, that is what I disliked about it the most. For example, when a student is given the problem: 456 divided by 12. Most of the time, the students are told to find out how many times 12 goes into 45. I agree that is the first step but sometimes that can take quite a long time. I know that I started by going 12 x 1, then 12 x 2 and etc. and had those problems written all over the page until I finally reached 12 x 4 which was too big! Even though I knew the first fact was 12, I still had 3 more problems to work out before I even found out the first part of the problem.

Then after subtracting and coming up with the value of 96 one must start the guessing process all over again. I don't know about you and your students but I know that many times doing the multiplication process over and over again all over the page causes a big mess. So I have found that instead of discouraging students with this process, I start every long division problem the exact same way which I will explain in the following paragraph.

To start the previously mentioned problem of 456 divided by 12, I would have students write down the multiples of 12 in a line down the side of their paper. The way I would have them do this is to write down the number 12 and circle it. Underneath it add another 12 which would be 24 and circle it. Since 24 is the second circle that means it is 12 times 2 (for the second circle). Continue with this process until the ninth circle is on the page.

Now lets start the problem using the chart that we just wrote. As one goes down the list, the first number that is the closest to 45 without going over is 36. So write 36 under the 45 and since 36 is the third circle put a 3 over the 5. After subtracting 45 - 36 to get 9 and bringing down the 6, now look for the number closest to 96 without going over which is exactly 96 and it is in the eighth circle down. Now put an 8 over the six and when you subtract there will be no remainder.

The great thing about this process is that it will work for ALL long division problems. It makes no difference what number you are dividing by and if it has a remainder or not. Students seem to like when they can be shown one way to work a problem especially when it works 100% of the time.

Lets switch to another area of mathematics being that called the Properties of Mathematics. These include the Commutative Property, Associative Property and Property of Identity. This is another area where the dictionary becomes a very useful tool. Divide the students into three groups and have the first group look up the word "commute", the second group look up "associate" and the last group look up "identical".

Starting with group one, have a student read the definition of the word commute. Have the students explain what they think it means in their world. The definition of the word means to drive back and forth. So the students should relate this to someone they know that drives from home to work and then after work is completed from work back to home. Now lets take their example using "H" for home and "W" for work and the "=" sign represents when work is done (just like in mathematics because after one has completed the working of a problem, the answer always follows an equal sign). Now the equation would look like this : H + W = W + H. Another suggestion with the "=" sign is that it looks like a road. The students might even relate to the equation more if they would use Home and School to represent themselves as being the commuters on a daily basis. Explain that the new equation would look like: H + S = S + H. This would show the students that they leave from Home every morning and go to school. When their work is done (=) at school they reverse the direction and travel from school back to home. During this time you can introduce the concept of distance being positive and that "subtraction" does not work for the commutative property.

To finish the concept of the commutative property one would switch to the way it is represented in mathematical terms. The topic of "substitution" (replacing a value with something of equal value) can be introduced to the students. Take the original equation of "H + W = W + H" and replace it with numbers. The students can select whatever values they want and then transform the equation into numerical values. For example the values might be "H = 4" and "W = 6". Their new equation would then look like 4 + 6 = 6 + 4. Clearly explain that the commutative property is nothing more than reversing the order of the numbers in the equation (just like the commuter reverses the order when returning home from work).

Secondly lets look at the concept of the "Associative Property". Once again I would use the dictionary to look up the definition of "associate". After the students have read the definition, I ask them if they ever associate with anyone in the cafeteria? Of course, we all know their answer. Next I take my hands and cup them around my mouth like I am going to tell someone a secret. When I pull my hands away from my mouth, I ask the students what mathematical symbol resembles my hands? Usually the students say, "Parenthesis". One of the key parts of the Associative Property is the moving of the parenthesis. Students seem to immediately make a connection of this symbol with the property.

To complete the explanation of this property, I usually refer to a commercial that I saw on TV. The commercial had a group of young people sitting in a cafeteria at a camp. The first youngster sitting at the end of the bench puts his hands around his mouth [his hands being parenthesis] and tells the second student on the bench. After that task has been completed [one will see the = sign] the second youngster puts his hands around his mouth to tell the secret to the third student. Now, it is important to stress to the students that these three youngsters NEVER moved but the only thing that did move was their HANDS. Now how does that transfer into a mathematical problem. Let's say that the first youngster is 1, the second youngster is 2 and the third one is 3. The equation would then look like this:

(1 + 2) + 3 = 1 + (2 + 3). The three youngsters are still sitting on the bench in the same position and the only difference is the movement of the parenthesis which are the youngsters hands. The parenthesis are around 1 and 2 on the left side of the problem. Once that is completed [the equal sign], then on the right side the parenthesis move around the 2 and 3 because the action is now taking place between these two youngsters. Once again, make sure and stress that the youngsters do not move while the action is taking place but the only thing that moves is their hands (or parenthesis). This is also a mathematical version of the "Telephone Game" that many kids have already played sometime before in their lives.

The final property to discuss is the Identity Property. This property can be used with addition or multiplication. During the introduction of this book, I briefly explained how to use the dictionary to look up the word identical. The key concept to instill in the students is that whatever value the problem starts out with, it must also be at the end of the problem (after the = sign). Examples: $5 + 0 = 5$ with the Identity element being ZERO and then using it for multiplication would be like $7 X 1 = 7$ with ONE being the Identity element. Stress to the students that the key idea of this property is to begin and end the problem with the same identical value.

A final concept that is important during the early stages of these grades is that of "Less than" or "Greater than". Students in the past of course have been taught one of the following: PacMan eats the bigger value and/or the alligator eats the bigger value. These are of course great techniques that will assist students with these problems but many students are still confused with the topic. I want to share an idea of how to relate this topic to an idea taught during the elementary years of teaching LEFT hand and Right hand. Remember having the students hold up each hand and whichever one made the "L" was the LEFT hand. Well, have the students do the same action now to decided which one is the LESS than or GREATER than sign. If you have the students move the Pointer finger towards their thumb, the

LEFT hand will make the LESS than sign. Once again tying together a previously taught concept with a math concept. Hopefully this will help students not only be able to correctly answer these questions, but be able to read them correctly as well.

<u>Individual Notes</u>

Individual Notes

Individual Notes

<u>Individual Notes</u>

Chapter 4

As students start into the middle school years many changes start to take place in the area of mathematics. Some schools start providing students with various levels of mathematics within a certain grade, the transferring into a higher level of mathematics and of course in many cases the emergence of the calculator.

Throughout this chapter, topics that are directly related to students in grades 6-8 will be addressed. Since these are the preparation years for their high school years, some of the topics will be seen again in the following chapter.

The first topic that should be introduced at this level will be "Mrs. Fenner's Golden Rule". This rule will help the student solve many types of problems from the sixth grade through their Senior year. It is a very simple concept and students will become amazed how many times the Rule can be applied. For example this rule can be applied to solving basic equations for a variable, finding the percent of given numbers and even simplifying problems in order to graph linear equations. As these topics are discussed throughout this chapter, one will quickly notice how useful this "Golden Rule" will apply in various ways.

The second topic to be discussed will be that of PERCENTS. Even though this topic is introduced at an earlier level, many times students are introduced how to change percents from fractions and decimals. At this level, the importance now focuses on the implementing of percents into various types of problems that students will not only see in mathematics but also in science and history.

For students to be able to correctly use percents, they must understand the definition of the word itself. The definition means "per one hundred". I usually tell my students that it is easy to remember the definition because when you look at the symbol (%) and break it apart, one can see the number "100". (The slash through the middle represents the "1" and the two circles of the symbol represents the two "zeros" of the 100.) This explanation really seems to stay in the minds of students because once again it transfers the mathematical definition into a visual image.

Students need to be able to convert fractions to decimals and decimals to percents and also reverse the process if necessary. To start the explanation of this process, it is important to review the concept of the hundredths place. Once again by looking at the % symbol, one can explain that since the symbol has two matching parts everything is going to deal with moving the decimal TWO places and TWO zeros in the bottom of the fraction. The "is over of" formula that so many students struggle with will be replaced by utilizing the "Golden Rule".

Another concept that I like to discuss with students is that of converting values into fractions, decimals and percents. Throughout the years, it has been my observation that many students struggle with various areas in the conversion process. Most understand how to convert the fraction to a decimal by using the division process of the numerator (top number) divided by the denominator (bottom number). But when it comes to converting from decimals to percents or vice versa, this is when problems arise. I have created a visual way for students to remember which way to move the decimal point. Previously it was explained that every concept dealing with percents involved the value of 2. So when performing this conversion, the decimal point must be moved either two places to the Left or Right. The simple method for students to remember which way the decimal point moves is by looking at the question. If the question says "Find the decimaL" the "L" at the end of the word clearly shows the student to move the decimal point Left. If the question says "Find the peRcent", the "R" in the middle of the word shows to move it Right. Even though this has no mathematical reasoning behind the process of Left or Right, students visualize the direction that they need to move the decimal point.

Third in this chapter, the topic of solving and simplifying basic algebraic equations will be discussed. These will include solving for a variable, simplifying an equation to isolate a given variable and the use of the substitution property with various equations. Once again the "Golden Rule" will play a major role in explaining the most direct and simplified means of solving these various equations.

Now the explanation of "Mrs. Fenner's Golden Rule". There are basically two steps to the rule and are easy for students to remember. The following are the steps for students to use: 1. The number on the side of the equal sign by itself must go in the calculator first. It is not the number listed first, the largest number or the one always on the right hand side of the equal sign but it is ALWAYS the number all alone. Now what that means is this special number will be the one without numbers, symbols, variables or values on the same side. For example in this question: $5X + 2 = 12$, the 12 is the number that is ALL BY ITSELF so it is entered into the calculator first. If you are not using calculators yet, tell the students that the number by itself is the one that is written down first on their paper.

2. The student then must DO the OPPOSITE of what the problem says. I always joke around with the students and say that this step should be easy for them since most of them are used to doing the opposite of what their parents say to do. At this time I always ask the students once again about synonyms and antonyms because I want them to know the opposite of various mathematical functions. I always have the students construct a chart by dividing a piece of paper into four sections. In each one of the individual sections, I have them list the terms; addition, subtraction, multiplication and division. Utilizing cooperative learning, I have the students divided into smaller groups and have them brainstorm about the words that are synonyms and antonyms for each of these words. It is

amazing the list that will emerge from this exercise. This is an important activity to do with your students. As a teacher, you can play this in a game type situation and reward the group that develops the largest list of synonyms and antonyms. If you are a parent, it is just as important to complete this exercise with your child.

Now I will step you through some problems using the "Golden Rule". First of all I will just start out with a simple one step basic algebraic equation: X + 5 = 12. Ask the student which number is on one side of the equal sign all by itself? Hopefully the students will answer by saying "12". It cannot be the 5 because there is an "X" and "+" on the same side as the 5. So have each student take their calculator and have them enter the number 12 (or have them write that number down first). Then ask the students what the left side of the equation says to do? The answer you are looking for is that the left side of the equation states to add 5 to X. Next it is important for the students to realize what the opposite of that would be in mathematics. The answer being searched for is to "subtract 5". So have the students complete that step in the calculator (or on paper) and the answer will appear. Even though this problem has an answer that students probably at this age would know prior to following the two steps, it is important for the students to understand the two step process. This also provides an opportunity for students to practice the "Substitution Process" by replacing the variable with their answer to see if it makes the equation true.

After completing the first problem more variations of the problems involving positive and negative numbers, all four of the basic functions, fractions and decimals should be given to the students. Most of all make sure to vary the location of the number that is to be entered into the calculator first.

This next example will utilize the "Golden Rule" more than once in order to demonstrate that no matter the complexity of the problem the rule will still work. The problem is 3x - 2 = 13. According to the "Golden Rule", ask the student to state which value must be entered into the calculator (or written on their paper) first and why. I cannot stress the importance of students being able to verbalize the steps as well as performing them correctly. Hopefully the student will respond by saying that the 13 will be entered into the calculator first (or written down first) because it is by itself. The next part of the rule says to do the opposite of what the problem says so your desired response is to ADD 2. Make sure that the equal sign is entered every time after completing a step to end the order of operations (or if doing it on paper make sure that they solve each step before moving on to the next step). The problem on paper now looks like this: $3x = 15$. Repeat the concept of what number goes in the calculator first (or written down first) and why. Hopefully the response is 15 and then ask them what opposite function would be used to solve the problem. This time it would be to DIVIDE by 3 and the answer would be 5.

The "Golden Rule" is very important for students to learn and understand because as they continue their journey throughout mathematics they will be able to use it quite often.

Individual Notes

Individual Notes

Individual Notes

Chapter 5

Moving on to the grade levels of nine through twelve, this chapter will not only open up new avenues to the students in this age group but will also tie together all of the information from their previous years together. The goal is at this age in their lives that the student will realize that math is a subject full of vertical growth. Throughout the chapter you will read about many topics that have been discussed earlier but it is important to reteach these topics at this level because I believe that the Freshmen year is very crucial for their future success.

(Also I want to repeat for those who only teach these levels and chose only to read this particular one!) It is important to realize that from the very beginning it essential for students to review and retain previous information as well as to gain new knowledge.

To review, I am going to start with the importance of the ABC's of mathematics. Previously, I discussed many areas of mathematics that students can learn about by putting them in alphabetical order. These areas include: base x height, length x width, the coordinate plane (X,Y) and the Pythagorean Theorem. At the high school level we can now include the trigonometric functions. After explaining the three sides of the right triangle (adjacent, opposite and hypotenuse) one can use the ABC's to teach two of the three trigonometric formulas which are COS and SIN. Explain that the common side used in both equations is the HYPOTENUSE because its location never changes in the right triangle. Its value will always fill the denominator of the fraction.

Now how can one create the trigonometric equations using the alphabet? Put the two beginning functions which are COS and SIN in alphabetical order. The two remaining sides are ADJACENT and OPPOSITE since we have already established that the HYPOTENUSE is in the denominator. When the COS and SIN functions are in ABC order, the two remaining sides (ADJACENT AND OPPOSITE) need to be in ABC order and become the numerator of those two functions. EXAMPLE: COS = ADJ/HYP. COS is first in ABC order and so is ADJ. The TAN equation unfortunately does not work with the ABC trick.

I believe it is important that students understand that many problems can revert back to the previously mentioned "Mrs Fenner's Golden Rule". If your students have not been introduced to this concept, it is important that it be taught before starting the year. At these levels, students can use this rule to not only solve basic equations but for also finding points on a line, solving equations with variables on both sides, percent equations, determining if lines are parallel or perpendicular as well as putting lines in "y=mx + b" order. Students will truly utilize this rule throughout the remainder of their Mathematical career.

Another concept that students struggle with is the parts of the Real Number System. Let's start this process by explaining the five main parts of the system which include: Irrational Numbers, Rational Numbers, Integers, Whole Numbers and Natural Numbers. The dictionary can assist with this process. To start, the students will look up the two words rational and irrational. Students will discover that these two words are antonyms of each other. The word "rational" refers to something that "makes sense or is realistic". Of course the word "irrational" refers to something that "does not make sense or is unrealistic".

Now discuss how those definitions apply to the world of mathematics. For example, rational numbers would include 3, -7, 1/2, 5 squared, 8.25, etc. These could all be represented through various ways. For example: on the calculator, with objects, using money, etc. The values are "real" or "make sense" to the students. Irrational numbers on the other hand would include: the value of "pi", the square root of 8, 5.249675…., etc. These numbers do not represent a "real value" or do "not make sense" to the students.

To make this concept clearer to students, I have written the following rules to utilize. Rational numbers must: 1. "Make Sense" and 2. You can predict the next number that would appear on the calculator screen. Irrational numbers on the other hand must: 1. "Not Make Sense" and 2. You cannot predict the next number on the calculator screen.

Now let's look at the specific parts located under Rational Numbers. The first one is that of INtegers. I realize that you are probably thinking that there was a typing mistake but actually I typed it that way on purpose. By typing the word "INteger", the definition will start by showing that it "INcludes" the "Negatives" to the students. The other important concept to stress to the students is that this group doesn't "INclude" decimals.

The "whOle" numbers are written so that students will recognize that this set of numbers start with the value of "zero". When written this way students can visualize the "O" in the word and easily remember the definition. Also when you say the word "whOle", students can see the formation of the zero with your mouth. Since they are also considered "INtegers", they cannot be decimals.

The remaining group of numbers is the "natura1". Once again the way the word is written provides a clue to the definition of a "natura1" number. The "natura1" numbers are a set of numbers that begin with the number "1". It is also the number that students "Natura11y" start counting with in Kindergarten. Since this set of numbers are also considered to be "INtegers" and "whOle", they do not contain decimals either.

Many times students are asked to identify the category or categories that a specific number would be included in meaning rational, integer, whole and/or natural. The chart I have designed will assist the students in answering this type of question.

REAL NUMBER SYSTEM

Rational   Irrational

INtegers

whOle

natura1

To use the chart above, explain to the students that on the left side that all natural numbers will be included in every group above it. The same is true with the whOle numbers and INtegers.

Now how does this help them with answering questions like: "What category/categories would the number "-3" be included in?". Have the students start at the bottom left side of the chart and find the first category -3 would be included in. It would not be "natura1" because they start with 1 and it would not be whOle because they start with zero. So since it starts in the INteger category, the chart tells the student that it is also Rational.

Now let's start with the number "12". Start at the bottom and ask the student is the number 12 a "natura1" number? If they answer correctly which would be "Yes", then the chart would tell them that 12 would be included in every group on the left side of the chart.

A set of properties that should also be reviewed at this age is that of the Commutative, Associative, Identity and Substitution. If the students are unsure of or completely forget these properties, refer back to Chapter 3 and discuss them again. These properties are important to students of this age especially it they are comparing their answers with other students or even with BOB (Back of Book).

Next I want to explore some of the concepts that I have discovered concerning the Coordinate Plane and using it to correctly graph lines. First of all the Coordinate Plane is divided into four sections known as the Quadrants. Take time to explain that the prefix "quad" means "four". The numbering of these quadrants may seem unusual but if one relates it to the concept of the sundial that was used by early Mathematicians it is much easier to understand. Start by labeling the axis like a compass using East, North, West and South beginning at the positive "x" axis and moving counter-clockwise. Then continue with the concept that the sun rises in the East and goes down in the West. So the quadrants are numbered 1 - 4 starting in the East.

Another important idea is that many students confuse the idea of which is the "x" and "y" axis. To make sure that this doesn't happen to your students, draw the "y" like this

The tail of the "y" will help the students remember which way that axis travels on the coordinate plane. Little subliminal hints I have found really help students. To continue with this idea, one can use the ABC concept to correctly teach graphing because ordered pairs are also in alphabetical order (X, Y).

One of the most important concepts that students must master throughout their mathematical career is that of slope and graphing lines. It is important for the students to realize that no matter how many lines one draws that there are only four different slopes possible. Slope is the direction and angle that a line is traveling.

A slope can take on an infinite amount of many values but the four directions include positive, negative, zero and undefined. The best way I have found for students to understand the concept of "Slope" is to relate it to something that they are familiar with in their everyday life. I relate this topic to driving a car. To start, I explain that the driver is crossing the United States. The driver is going from West to East.

While crossing the United States, the driver will find out that his "Feet Above Sea Level" will change. There are four possible changes in his direction which include an increase (positive slope), a decrease (negative slope), no change (zero slope) or finally it is impossible for him to drive straight up/down the side of a mountain (undefined slope).

SLOPE CHART

Positive

Negative

Zero

Undefined

Once the students understand the "Slope Concept", it is important to start the teaching of the equation "y=mx+b". Most students are instructed what each part of the equation represents but many never understand how it applies to the process of graphing a line. Yes, I always start by explaining that the (x,y) represent any point that will be on that line. There are an infinite amount of points on that line so these two letters will remain variables since they are constantly changing.

The "m" represents the slope of the line. This value will become a constant in the equation because no matter what location you select along the line the slope will remain the same. At this point I also introduce two important words that begin with the letter "m" that will assist the students with understanding the total slope concept. Those two words are "middle" and "move". "Middle" tells the student where the slope value is located in the equation "y=mx+b". "Move" on the other hand tells the student what direction the line is moving.

The "b" represent the "y-intercept" (where it crosses the y-axis) for the line. This value will also become a constant because the line will only cross the y-axis in one place. At this point, I will introduce two words beginning with the letter "b" that will assist the students understanding the y-intercept concept. These two words are "Back" and "Beginning". "Back" tells the students where the y-intercept is located in the equation. The word "Beginning" tells the students that is the point that they use to BEGIN graphing the line.

Students will start with equations that are already in the correct form of "y=mx+b". I begin by basically asking them for the "slope" and "y=intercept" and then ask what those two values tell them about the line. I have them to just sketch a basic drawing of a line to demonstrate where the line crosses the y-axis and which direction it will be moving. This process allows the students to start visualizing the graph of a line prior to drawing it. This process also reminds the students that these two values are constants no matter at what location they are on the line.

Secondly, the students will start drawing the graphs accurately by counting the slope. Still emphasize that they must "Begin" with the "b" value located at the "Back" of the problem and then use the "m" value which is located in the "Middle" to instruct them of which way to "Move". It is amazing how much easier students will remember the "m" and "b" values and their functions by using these other words.

Of course there will always be times when the equation given to the students to graph will not be in the "y=mx+b" form. I then introduce the concept of "Switch Sides so Switch Signs". This means if the value must "switch sides" of the equal sign then it must also "switch signs". For example: 2x = -y + 4. The "x" and "y" values are on the wrong side of the equal sign to be in "y=mx+b" form. Since they must "switch sides" they must "switch signs". So after "switching" the values, the equation will now become y= -2x+4. As one can see, the "y'" is now positive and the "2x" became negative.

Of course one will surely guess that "Mrs. Fenner's Golden Rule" must relate to the topic of graphing lines. How does that two step rule apply to graphing lines?

Many times students are given an equation of a line with the value of "m" or "b" missing and asked to find it by using a point on the line. Example: Given the equation $y=3x + b$ and the point (-2, 4) is on the line, find the value of "b". First of all instruct the students to use the process of substitution and get the equation of $4=3(-2) + b$. Simplify the equation to $4= -6+b$. Now comes the "Golden Rule". The number 4 goes in the calculator first and add 6 (do the opposite). The final answer is $10=b$.

The final concept to discuss in this chapter is the Unit Circle. Throughout my years of teaching Trigonometry, the Unit Circle is a concept that is one that is important for students to understand how it was designed and how it relates to the Trigonometric functions that they learned in Algebra. Once the Unit Circle is drawn, the trigonometric functions are added at the degrees of 30, 45, 60, 90 etc. The (x,y) coordinates as we discussed earlier are in alphabetical order and now on the Unit Circle. The trigonometric values are also in ABC order by being in the order of (cos, sin). It is amazing how many times that alphabetical order plays a significant role in mathematics.

After converting the degrees to radian measure, it becomes quite easy for students to remember the (cos, sin) values without a calculator. Look at the chart below and I will show you how students can quickly learn the trigonometric values with radian measure.

CHART

30° = π/6   (√3/2, 1/2)

45° = π/4   (√2/2, √2/2)

60° = π/3   (1/2, √3/2)

Now if you look at the denominator of the π/6, the value of the first number in the ( ) is √3/2. If you multiply 3 x 2 you come up with the 6. If you look at the denominator of π/4, the value of the first number in the ( ) is √2/2. If you multiply 2 x 2 or add 2+2, you come us with the 4. Lastly if you look at the denominator of π/3, the first number in the ( ) is 1/2. If you add 1+2, you will come up with the 3. This may seem a little extreme but students always like it.

Individual Notes

Individual Notes

Individual Notes

Chapter 6

In this chapter of the book, I would like to address a major problem that almost every math student faces through their mathematical career. That problem would be that of becoming successful solving word problems. Most of the time, many math students will quickly admit that the main reason they fail at this type of problem is because they do not like to take the time to read the problem. So the process that I have developed to provide a path for students to excel at word problems will always provide them with a starting point.

The process I have devised to help students with Word Problems is called the "AIMS" process. I designed the acronym for several reasons. I had an "AIM" for designing this process which is to change students' attitude towards these problems. I also created this acronym so that it provides the order the students must follow to solve these problems. Finally each letter will instruct the students of what they are supposed to do in each of the four steps.

The first letter "A" represents "ACTION". The students need to answer the question of "What ACTIONS (verbs) are listed in the problem for ME to complete?". These verbs are usually found at the END of the problem. One needs to stress that these are not the actions the individuals in the problem are doing but what the problem solver needs to do. Some examples could include: Find the area of the triangle; Solve how much money each person will need.; etc.

The second letter "I" represents the word "INFORMATION". The student needs to record all types of information that is found within the problem. Examples: $45 for the price of one sweater, four students received an "A", 20% of the girls picked a green pencil, etc.

The third letter "M" represents the word "MATHEMATICS/METHODS". This is the step where the students need to decide what mathematical processes are needed to use to find the solutions for the "ACTIONS".

The final letter "S" represents the word "SOLUTION". Many times students may find the answer but never write down the final solution on the answer document. Also this is an opportunity for the student to practice the "restate" process used in their English/Literature courses. Since creating this process, I have had my students write the answer in a sentence form to not only practice restating the problem but to also help them realize if they answered the original question.

I have also created a layout that the students use to follow this method. It will also provide the students with a neat and orderly process so that the student and person grading the question can follow as well.

First I have the students draw a capital letter "I" on the paper.

Above the top line, the students write the letter "A" and that is where the students are to list all of the actions that they are to take in order to find the answer. Below that line on the left, the students need to write the letter "I". In this section, the students need to write down all of the information found within the problem. On the right side, have the students write the letter "M". In this section the student needs to write all of their mathematical calculations. To finish the problem the students need to write the letter "S" below the bottom line. In this section, the student is to state the answer/answers to the problem. See the Example Problem and Student Work on the next 2 pages.

Three different opinion polls show different results for the proportion of voters expected to vote for Candidate A in an election for mayor.

Poll 1: Nine of every 20 voters are expected to vote for Candidate A.

Poll 2: The percentage of voters expected to vote for Candidate A is 52%.

Poll 3: There are 130,000 people expected to vote, and of these, 55,000 are expected to vote for Candidate A.

In your Answer Document, determine which of these polls shows the greatest favorable result for Candidate A. Show your work to provide an explanation for your answer.

The following page clearly illustrates the use of the AIMS Process on the previous problem. This would be an example to use with your students.

AIMS

A 1. Determine the Best Poll   2. Show your work

I	**M**
1. Poll 1 9/20	1. 9 divided by 20 = .45 or 45%
2. Poll 2 52%	2. 52%
3. Poll 3 55,000/130,000	3. 55 divided by 130 = .42 or 42%

S

The Best Poll is Poll 2.

A ction (Verbs)
I nformation
M ath/Method
S olution

Designed by Terri Wolfe Fenner, N2LearningNow.com

<u>Conclusion</u>

At this point of the book, I hope that you have read it from beginning to end. It is the only way you can truly visualize the vertical pathway of mathematics. Many ideas may seem very extreme to you, but one of the ways to get students attention is to demonstrate creativity. Anyone can repeat what is written in a textbook. Anyone can copy worksheets from workbook. But once students realize that there are numerous ways to work the same problem and they can pick the way they understand, math becomes an individual academic area.

Another way to have students buy into the area of mathematics is to demonstrate that you work as hard as they do to try and understand it. Some of the History of Mathematics that I incorporate into my lessons also lead to pulling the students attention into the lesson. If the students feel that you are walking textbook produced by someone else, they tend to become a boring textbook as well.

Make students believe in you and your ideas. That is what they will remember in the years to come as well as the math that you taught them. It is exciting when students I taught in the past tell me that "I still use Mrs. Fenner's Golden Rule" today.

Make sure that you look for the next book in the series. It will address other successful ways to address Word Problems as well as more "tricks" to help our children understand the World of Mathematics.

My contact information is:

Website: N2Learningnow.com

Email: terri@n2learningnow.com

www.ingramcontent.com/pod-product-compliance
Lightning Source LLC
Chambersburg PA
CBHW071720170526
45165CB00005B/2087